THE ABC'S OF DINOSAURS

A DINO-LOVERS JURASSIC EXPLORATION!

THE ABC'S OF
DINOSAURS
A DINO-LOVERS JURASSIC EXPLORATION!

Welcome to the Prehistoric World!

Imagine stepping into a world filled with towering giants, swift hunters, and incredible creatures unlike anything you've ever seen. In Dino Explorers: Fun Facts and Adventures with the Prehistoric Giants, you'll journey back in time to when dinosaurs ruled the Earth. These fascinating creatures came in all shapes and sizes—from tiny, feathered runners to colossal plant-eaters with long necks stretching to the sky. Some had armor, some had claws, and others had horns or sails that made them look like real-life superheroes.

This book is your guide to the amazing world of dinosaurs. Packed with fun facts, engaging descriptions, and a sprinkle of imagination, it brings these prehistoric wonders to life. You'll learn how they lived, what they ate, and how they thrived in their ancient habitats. Whether you're curious about fierce predators like the mighty T. rex or gentle giants like Apatosaurus, there's something here for every dinosaur lover.

So grab your explorer's hat, and let's embark on an adventure to uncover the mysteries of these incredible creatures. By the end of this book, you might even feel like a real paleontologist!

Allosaurus

Pronunciation: uh-LOH-sawr-us

Allosaurus was a super cool meat-eating dinosaur with sharp teeth and claws that loved to hunt for its meals. It lived a long time ago, during the Jurassic period, and it was one of the top predators of its time. Allosaurus had a big head with many teeth, making it a fearsome hunter, like a T. rex but smaller.

The Allosaurus's name means "different lizard. " It was different from other dinosaurs because of its size and features. Imagine meeting an Allosaurus—it would be like meeting a real-life dinosaur superhero!

Did You Know?
Allosaurus had excellent eyesight, which helped it spot prey from far away. This made it a master hunter in the dinosaur world!

Ankylosaurus

Pronunciation: an-ky-LOH-sawr-us

Ankylosaurus was like a giant tank with hard, bony armor that kept it safe from hungry predators. Its big tail could swing around to scare away enemies or even act as a strong weapon. Although this dino had a small brain, its strong body and tough armor kept it safe.

Ankylosaurus walked on four short legs and was a plant-eater, munching on leaves and plants. If you had an Ankylosaurus friend, you'd feel super safe with its armor and tail guarding you!

Did You Know?
Ankylosaurus had nostrils that led through a maze-like system in its skull, which may have helped it cool down or even create sounds!

Apatosaurus

Pronunciation: a-puh-TOH-sawr-us

Apatosaurus was a massive, long-necked dinosaur that loved to eat leaves from the tallest trees. Even though it was huge, it had a tiny head and swallowed rocks to help it digest tough plants. This dino had a whip-like tail that could make loud cracking sounds—kind of a big dinosaur whip!

Apatosaurus was such a big dino that its heart was probably as big as a small car! If Apatosaurus visited your town, it would need to bend its long neck to say hello through your window!

Did You Know?
The apatosaurus's long neck had air sacs in its bones, which made it lighter and easier to lift—like built-in dinosaur balloons!

Archaeopteryx

Pronunciation: ahr-kee-oh-PEH-triks

Archaeopteryx was a hybrid of a bird and a dinosaur—it had feathers and teeth! It was one of the first birds in the world, but it looked more like a small dinosaur with wings. This special dinosaur bird could fly a little but might have spent more time hopping and gliding.

Archaeopteryx had sharp claws on its wings, which helped it climb trees and catch insects. If you had a pet Archaeopteryx, you could teach it tricks with its feathers and wings!

Did You Know?
The first Archaeopteryx fossil was discovered in 1861, just one year after the famous scientist Charles Darwin published his theory of evolution!

Brachiosaurus

Pronunciation: brah-see-OH-sawr-us

Brachiosaurus was a supertall dinosaur with a long neck that helped it reach the highest leaves on trees. Its front legs were longer than its back legs, making it look as if they were always standing on tiptoes. This dino was so big that its footsteps could shake the ground like a mini earthquake!

Brachiosaurus was a peaceful plant-eater that had to eat a lot to keep its huge body fed. If Brachiosaurus was playing hide-and-seek, you could spot it from far away because of its tall neck peeking out!

Did You Know?

Brachiosaurus had a unique hump on its back that might have helped store extra food or water for long journeys!

Baryonyx

Pronunciation: bar-ee-ON-iks

Baryonyx was a fierce fish-eating dinosaur with long snouts and sharp teeth that were made for catching slippery prey. It was like a crocodile with long arms and claws to help it catch fish in rivers. This dino had a giant claw on its thumb, which might have been used for fishing or defending itself.

Baryonyx could swim and hunt in the water, making it a unique and fantastic dinosaur. Imagine if Baryonyx invited you for a fishing adventure – you'd have the most relaxed dino-fishing buddy!

Did You Know?
Baryonyx was one of the first dinosaurs discovered in Europe, and its fossils were found in England!

Bambiraptor

Pronunciation: bam-bee-RAP-tor

Bambiraptor might have been a small dinosaur, but it was quick and clever. Its name is funny because it sounds like "Bambi," but this dino wasn't a deer—it was a fast predator. It had sharp teeth and claws, which it used to catch small animals and insects.

This dino might have had more feathers than other dinosaurs. If Bambiraptor were in a race, it would win because of its speedy and sneaky skills!

Did You Know?
The bambiraptor had excellent vision and could see its prey even in low light, making it a master hunter at dawn and dusk!

Buitreraptor

Pronunciation: boo-it-reh-RAP-tor

Buitreraptor was a cool dinosaur that lived in South America and was a bit like a dino-bird. Its name means "vulture thief," which might be because it might have stolen food from other dinosaurs. Buitreraptor had long legs and a long tail, which helped it balance and run fast.

This dino had sharp claws on its toes for catching prey and climbing trees. If Buitreraptor was in a forest, you might see him jumping and showing his skills!

Did You Know?
Buitreraptor had feathers, just like modern birds, which helped it stay warm and maybe even attract friends!

Carnotaurus

Pronunciation: kar-noh-TAWR-us

Carnotaurus had tiny arms similar to T. rex, though even shorter. This dinosaur was a super-fast runner, using its strong legs to chase its prey. Its bumpy skin was covered in many small bumps, giving it a unique texture.

It had horns on its head above its eyes that made it look as a dino bull ready to charge. Imagine Carnotaurus racing against a car – it might have been faster than some race cars from today!

Did You Know?
Carnotaurus had a highly specialized skull that allowed it to move its jaws quickly, making it an efficient predator!

Ceratosaurus

Pronunciation: seh-RAT-oh-sawr-us

Ceratosaurus had an excellent nose horn and bony knobs on its head – like a dino with its built-in decorations! This dino was a stealthy hunter, sneaking up on other animals to catch them by surprise. Ceratosaurus had strong jaws and sharp teeth, perfect for munching on meat and snacks.

It might have lived in groups or families. And If Ceratosaurus was in a talent show, it could show off its awesome headgear and roar for the audience!

Did You Know?

Ceratosaurus had a unique sail-like structure on its back made from long spines, which might have helped it stay cool or attract friends!

Compsognathus

Pronunciation: com-pso-GNATH-us

Compsognathus was a pint-sized dino faster than a cheetah – it could give Usain Bolt a run for his money! With its sharp teeth and speedy legs, this dino loved to snack on insects, small animals, and probably even dino snacks. Compsognathus had a long tail that helped it balance, making it a pro at hopping and running.

It might have traveled together with other Compsognathus buddies, like a tiny dino party on the move. Imagine having Compsognathus as a pet – you'd have a super-fast and agile dino friend for all sorts of adventures!

Did You Know?
Compsognathus means "small jaw" in Greek because its jaw was much smaller than other dinosaurs!

Corythosaurus

Pronunciation: ko-RITH-oh-SAWR-us

Corythosaurus was a dinosaur singer with a special crest on its head that helped it make loud, musical calls. This dino loved munching on leaves and plants and was a dinosaur vegetarian chef. It walked on two legs, and its head crest made it look like a dinosaur rock star with funky hair.

It probably lived near rivers and swamps, wading in the water to find tasty water plants. Imagine Corythosaurus concerts in the dino forest – you'd be dancing to the dino beats and singing along!

Did You Know?

The name Corythosaurus comes from Greek words meaning "helmet lizard" because of its awesome crest on its head!

Diplodocus

Pronunciation: dip-LOH-doh-kus

Diplodocus was a giant dinosaur with a long neck and tail—like a dinosaur version of a stretch limo! This dinosaur had tiny teeth and probably ate plants, reaching high up in the trees with its long neck. Diplodocus might have whipped its tail to make loud sounds or scare off other dinosaurs.

It walked on four legs and was a gentle dino, not interested in chasing meat. If Diplodocus went to a dino party, it would have to bend down its long neck to talk to its shorter dino pals!

Did You Know?
Diplodocus had hollow bones, which made its massive body lighter and helped it move more easily despite its huge size!

Deinonychus

Pronunciation: dye-NON-ih-kus

Deinonychus was a clever dino with a fancy "terrible claw" on its foot, perfect for catching prey and giving high-fives (not really, though!). This dino liked teamwork and might have hunted in packs. Deinonychus was a speedy runner and a good jumper, making him a dino athlete in his own right.

It had sharp teeth for munching on meat, and its brain was more significant than some other dinos. Imagine having Deinonychus as a buddy – you'd go on daring dino missions together, hunting and exploring!

Did You Know?
Deinonychus had feathers! Scientists believe it was one of the first dinosaurs to have feathers, making it look a bit big, fast bird.

Dilophosaurus

Pronunciation: di-LOH-FO-sawr-us

Dilophosaurus, a fancy name that means "double-crested lizard, " had two crests that looked like dino antennas on its head. This small but fierce predator might have spit venom to catch its prey from far away.

It was a small but fierce predator that loved snacking on meat. With sharp teeth and two legs for running, Dilophosaurus was very fast. Imagine if it performed in a talent show, showing off its crests and even doing a dino dance!

Did You Know?
Dilophosaurus was one of the earliest large predatory dinosaurs, living over 193 million years ago during the Early Jurassic period!

Dromaeosaurus

Pronunciation: droh-mee-oh-SAWR-us

Dromaeosaurus was a small dino with a long name and sharp claws on its feet! This quick and clever creature might have hunted in groups of dino ninjas. Its name means "running lizard," and it had sharp teeth for catching meat.

It had a curved, sickle-like claw on its foot that helped it catch prey and maybe even climb trees. Imagine Dromaeosaurus joining your dino adventure – you'd have an agile and intelligent buddy for all sorts of daring quests!

Did You Know?
Dromaeosaurus is believed to have had feathers, just like modern birds. These feathers helped it stay warm and may have aided in hunting and balancing while running!

Edmontosaurus

Pronunciation: ed-MON-toh-SAWR-us

Edmontosaurus had a toothless beak resembling a duck's bill and loved munching on yummy plants—it was a giant, salad-loving dinosaur! This dino had a paddle-shaped tail that probably splashed water at dino pool parties. With bumpy skin and a tendency to travel in big groups that formed a dino parade,

Edmontosaurus was quite a sight. It grew quickly, reaching its full size faster than many other dinos—a true dino rocket. Imagine Edmontosaurus dancing in the rain with its tail, putting on a dino water show!

Did You Know?
The name Edmontosaurus comes from Edmonton, a city in Canada where some of its first fossils were discovered!

Einiosaurus

Pronunciation: ee-NEE-oh-SAWR-us

Einiosaurus had a cool name that means "buffalo lizard" – it probably had a nose horn, like a fancy dino bull. This dino might have used its horn to battle other dinosaurs, acting as a brave knight defending its territory.

As a plant-eater, Einiosaurus munched on leaves and plants, serving as the gardener of its prehistoric world. It had a bony frill on its head that protected its neck against sneak attacks. Einiosaurus joined a dino parade, and it would be the star with its impressive horn and fancy frill!

Did You Know?
Einiosaurus lived during the Late Cretaceous period, and its fossils were discovered in North America, specifically in Utah!

Eoraptor

Pronunciation: ee-oh-RAP-tor

Eoraptor was one of the earliest dinosaurs from the prehistoric Dino family album. It was small and walked on two legs, making it a pioneer in the world of bipedal dinosaurs. Eoraptors had sharp teeth and were probably meat-eaters.

This dino might have looked like a tiny T. rex with its sharp teeth and fierce demeanor. Imagine traveling back in time to meet Eoraptor – you will see one of the very first dinosaurs ever!

Did You Know?
Eoraptor fossils were discovered in Argentina, making it one of the first dinosaurs found in South America!

Euoplocephalus

Pronunciation: yoo-uh-PLOH-suh-FAY-lus

Euoplocephalus was a tough dino with bony plates on its back and a club-like tail, ready for dino battles. This dino was a plant-eater, munching on leaves and bushes. Euoplocephalus had spikes to keep predators away.

It also walked on four legs and was a bit of a dinosaur tank, slowly moving around with its armor. Imagine if Euoplocephalus was a dinosaur superhero—it would have an invincible tail and armored back to save the day!

Did You Know?
Euoplocephalus had a highly specialized armor of over 500 bones, including spikes and plates, making it one of the most heavily armored dinosaurs ever discovered!

Fabrosaurus

Pronunciation: FAY-broh-SAWR-us

Fabrosaurus was a tiny dinosaur that lived a long time ago and loved munching on plants. Even though it was small, it had sharp teeth for nibbling on leaves and tasty dino snacks. This dino was a little chef, using its teeth to prepare plant-based meals.

Fabrosaurus walked on two legs and had a tail for balance, like a dino acrobat. Imagine having Fabrosaurus as your tiny dino friend – you'd go on mini adventures exploring the dino world together!

Did You Know?
Fabrosaurus is believed to be one of the earliest known dinosaurs, making it a key species for scientists studying dinosaur evolution.

Fukuiraptor

Pronunciation: foo-koo-ih-RAP-tor

Fukuiraptor was a speedy dinosaur with sharp claws on its feet, perfect for catching prey on the run. Its long tail helped it balance and move gracefully. Fukuiraptor was a meat-eater who chased after other animals and showed off their hunting skills.

It had sharp teeth to munch on meat and probably enjoyed a dino buffet of snacks. Imagine Fukuiraptor in a dinosaur race—it would be like watching a fast and fierce dinosaur Olympic event!

Did You Know?
Fukuiraptor was discovered in Japan's Fukui Prefecture, making it one of the few dinosaurs from the Early Cretaceous period found in Asia!

Falcarius

Pronunciation: fal-KAIR-ee-usr

Falcarius was a plant-eating dino that loved feasting on leaves and other green goodies – a real dino herbivore! It had sharp claws on its hands, but it wasn't interested in meat – just using them for a plant buffet! Falcarius had a unique mix of features – it walked on two legs and used its arms to grab food.

This dinosaur lived in what is now North America and probably hung out with other dinosaurs in groups. Imagine if Falcarius hosted a dinosaur picnic—it would prepare the best veggie dishes for its dinosaur friends to enjoy!

Did You Know?

Falcarius is one of the earliest known therizinosaurus dinosaurs, showing a special mix of features that helped scientists understand how some dinosaurs evolved from meat-eaters to plant-eaters!

Gallimimus

Pronunciation: gal-ih-MY-mus

Gallimimus was a dino sprinter with long legs, making it one of the fastest runners in dino-land! It had a funny beak and ate plants, but it might have also snacked on small critters. Gallimimus had many sharp teeth, like a dino smile full of tiny razors.

It might have traveled in groups, running together in a dino parade. Imagine a Gallimimus race – as watching a Dino Olympics with these speedy athletes!

Did You Know?
Gallimimus could run up to speeds of 35 miles per hour, making it one of the fastest dinosaurs ever discovered!

Giganotosaurus

Pronunciation: jye-guh-NOH-tuh-SAWR-us

Giganotosaurus was a ginormous meat-eating dino whose name sounds like it's from a dino superhero movie! This dino was a bit similar to T. rex but lived in a different part of the world, being a "terrible lizard" of its own. Giganotosaurus had sharp teeth that could crunch through bones, making it a fierce predator.

It walked on two powerful legs and might have hunted in groups of dino hunters. If Giganotosaurus joined a dino talent show, it would win the "biggest dino" prize, hands down!

Did You Know?

Giganotosaurus is one of the largest known terrestrial carnivores, even bigger than the famous T. rex, measuring up to 40 feet long!

Giraffatitan

Pronunciation: jee-RAF-uh-TIE-tan

The giraffatitan was a dino giraffe with a super long neck that helped it reach tasty leaves high up in the trees. This dino was a plant-eater, munching on green stuff and probably keeping the dino forest neat and tidy. The giraffatitan was so big that its heart was probably the size of a small car – a real dino giant!

It walked on four sturdy legs and had a long tail, like a dino with its built-in balancing stick. Imagine if Giraffatitan was a dino firefighter – it could use its long neck to extinguish dino forest fires!

Did You Know?
Giraffatitan was first discovered during the famous Tendaguru Expedition in Tanzania, which unearthed some of the largest dinosaur fossils ever found!

Gorgosaurus

Pronunciation: gor-go-SAWR-us

Gorgosaurus was a mean, lean, meat-eating machine with sharp teeth and a powerful bite. It was T. rex's cousin—a dinosaur predator with a "dino king" attitude. Gorgosaurus might have hunted in groups, like a pack of wolves looking for a tasty dinner.

It had tiny arms and was a fast runner, making it a dino chaser in its prime. If Gorgosaurus had a dino dance party, you'd see its sharp teeth shining as it rocked out to dino tunes!

Did You Know?
Gorgosaurus was first discovered in Alberta, Canada, and its fossils have provided scientists with valuable insights into the lives of large tyrannosaur dinosaurs!

Hadrosaurus

Pronunciation: ha-DROH-sawr-us

Hadrosaurus was a dino with a funny-looking duck-like beak and loved munching on plants. This dino had rows of teeth, and if one tooth got worn out, another took its place! Hadrosaurus was a good walker and swimmer – it could be a dino lifeguard if it wanted!

It lived a long time ago in what's now North America, hanging out with other dinos and having dino parties. Imagine Hadrosaurus singing dino songs in the dino choir – its beak would be a dino microphone!

Did You Know?
Hadrosaurus is often called the "duck-billed dinosaur" because its beak looked just like a duck's, helping it to efficiently gather and eat a variety of plants!

Herrerasaurus

Pronunciation: her-REHR-uh-SAWR-us

Herrerasaurus was one of the earliest dinosaurs, like a dinosaur time traveler from long ago. It was small but tough, with sharp teeth and claws that made it a fearsome hunter. Herrerasaurus acting as a dinosaur detective—it lived in what's now South America and left clues about early dinosaur life.

It walked on two legs and might have even been a fast runner, chasing after prey in the dino wild. Imagine Herrerasaurus solving dino mysteries in its dino trench coat and hat – a true dino detective!

Did You Know?
Herrerasaurus is one of the oldest known dinosaurs, dating back to about 231 million years ago during the Late Triassic period, helping scientists understand how dinosaurs first evolved!

Hesperornis

Pronunciation: hes-pur-OR-nis

Hesperornis was a dinosaur that was great at swimming and diving—it was like an Olympic swimmer! This dinosaur had strong legs and webbed feet that helped it paddle through the dinosaur waters. Hesperornis loved fish and had sharp teeth to catch its underwater meals.

It might have had a fun dino life in the lakes and oceans. And If Hesperornis had a dino pool party, it would be the champion of the diving contest with its dino flips!

Did You Know?
Hesperornis could hold its breath underwater for up to 2 minutes, making it an amazing underwater hunter and explorer!

Hypsilophodon

Pronunciation: hip-suh-LOH-fuh-don

Hypsilophodon was a speedy dino with long legs and a name that sounds like it's from a dino race track! This dino was a plant-eater, munching on leaves and plants to keep its dino energy up. Hypsilophodon might have lived in groups, running around together and having dino fun.

It had sharp teeth and a sharp beak, perfect for grabbing leaves and taking dino bites. Imagine Hypsilophodon in a dino marathon – it would be the winner with its super dino speed!

Did You Know?
Hypsilophodon fossils were first discovered in the famous dinosaur-rich Isle of Wight in England, making it one of the most well-known small dinosaurs from the Early Cretaceous period!

Iguanodon

Pronunciation: ig-wuh-NOH-don

Iguanodon had thumbs that looked like spikes, and it used them to munch on yummy plants – like a dino gardener! This dino had a bumpy thumb and fingers that it could use for grabbing leaves and branches. Iguanodon walked on two legs and sometimes on all fours, like a dino that knew different dance moves.

It had a long, curved thumb spike, which was a bit like a dino Swiss Army knife. Imagine Iguanodon at a dino picnic – it would be the chef, using its thumbs to skewer dino kebabs!

Did You Know?

Iguanodon footprints have been found in groups, showing that these dinos liked to walk and explore together, just like a big dinosaur family on an adventure!

Indosuchus

Pronunciation: in-DOH-soo-kus

Indosuchus was a meat-eating dinosaur with a name that sounded like a secret agent from dinosaur spy movies. This dino had sharp teeth and was a dinosaur detective hunting for its prey. Its bumpy skin resembled that of a dinosaur wearing armor, ready for action.

It was a dino lion with powerful jaws and sharp teeth for taking big dino bites. Imagine Indosuchus on a dino adventure – it would be the fearless dino leader solving dino mysteries!

Did You Know?
The name Indosuchus comes from "Indo," meaning India, and "suchus," meaning crocodile, because scientists thought it looked a bit like a crocodile!

Incisivosaurus

Pronunciation: in-sigh-sih-VAWR-us

Incisivosaurus was a fancy name that means "incisor lizard" because it had super long front teeth that looked like a dinosaur bunny's! This dinosaur loved eating plants and leaves, using its teeth to snip off tasty bites. It had a small body and a long tail, making it look similar to dino artists!

It lived in what's now China, hanging out with other dinosaurs and having dino parties. Imagine Incisivosaurus nibbling on dinosaur snacks— a dinosaur chef preparing a tasty meal!

Did You Know?
Incisivosaurus fossils were discovered in the famous Jehol Biota of China, a place known for preserving many amazing dinosaur fossils!

Irritator

Pronunciation: ih-ri-TAY-ter

Irritator had a funny name because when scientists found its skull, they were annoyed by the rocks covering it – like a dino puzzle! This dino was a meat-eater with sharp teeth, and its name became a cool reminder of the dino discovery. Irritators had a long snout and might have hunted for fish near dino rivers and lakes.

It was a dino crocodile with its sharp teeth and aquatic lifestyle. Imagine Irritator in a dino riddle game – its name would be the tricky clue everyone tries to solve!

Did You Know?

Irritator had special nostrils at the top of its snout, just like a modern crocodile, allowing it to breathe while partially submerged in water!

Janenschia

Pronunciation: ya-nen-SHEE-uh

Janenschia was a big dino with a long name and neck – like a dino stretch limo with a view! This dino was a plant-eater, using its long neck to munch on leaves from the tallest trees. Janenschia was a dino giraffe, reaching high for its tasty dino snacks.

It walked on four sturdy legs and was a peaceful dino, not interested in chasing after meat. Imagine Janenschia giving dino friends high-fives with its long neck – it would be a dino party trick!

Did You Know?

Janenschia was named after Werner Janensch, a famous German paleontologist who discovered many dinosaur fossils in Africa!

Jeholopterus

Pronunciation: juh-LOH-pter-us

Jeholopterus was a dino that could fly – it's like a dino pilot soaring through the dino skies! This dino had leathery wings, and it flew during the time of the dinos. Jeholopterus probably hunted for insects, catching them as a mid-air dino superhero.

It might have lived in what's now China, exploring the dinosaur world from above. Imagine Jeholopterus in a dinosaur air show—it would be the star performing loops and dives with its dinosaur wings!

Did You Know?

Jeholopterus fossils are some of the best-preserved pterosaur fossils ever found, including details of their wing membranes and even some soft tissues, helping scientists learn how these amazing dinos flew!

Jinfengopteryx

Pronunciation: jin-FENG-op-ter-iks

Jinfengopteryx was a dino with feathers that lived long ago – it's like a dino with a feathery coat! This dino was a small dino-bird with its feathers and beak. Jinfengopteryx had sharp teeth, so it was probably a meat-eater, catching small animals and insects.

It had a long tail and might have been a good climber. Imagine Jinfengopteryx in a dinosaur fashion show—it would be the star model showing off its stylish feathered look!

Did You Know?
Jinfengopteryx is named after a mythical Chinese bird called Jinfeng, or "golden phoenix," because of its beautiful feathered look!

Juravenator

Pronunciation: joo-rah-VEN-ay-tor

Juravenator was a small dino buddy you could carry in your backpack on adventures! It had feathers that made it similar to a dino bird, even though it was a meat-eating dino. Juravenator might have hunted for insects and small animals, like a dino detective searching for snacks.

This dinosaur was found in Germany, giving us a peek into the dinosaur world from long ago. Imagine a Juravenator in a dinosaur dress-up party—its feathers would be a dinosaur costume!

Did You Know?

Juravenator fossils were so well-preserved that scientists could see tiny details of its skin, making it one of the best-preserved small dinosaurs ever discovered!

Kentrosaurus

Pronunciation: KEN-troh-SAW-rus

Kentrosaurus had spikes on its back and tail, making it look like a real-life dino superhero with built-in armor! This dino was a plant-eater, munching on leaves and green stuff. Kentrosaurus was a bit of a dino porcupine – its spikes could scare away dino bullies.

It lived in what's now Africa and might have hung out in groups. Imagine Kentrosaurus showing off its spikes in a dinosaur fashion show—it would be the star of the runway!

Did You Know?
Kentrosaurus had two long spikes on each shoulder, making it look extra fierce and well-protected against other dinosaurs!

Kosmoceratops

Pronunciation: kos-mo-SAIR-uh-tops

Kosmoceratops had a name that sounded from a dino space adventure – it's like a dino explorer from another planet! This dino had fancy frills on its head with cool patterns. Kosmoceratops was a plant-eater with sharp beak teeth for munching on tasty leaves.

It lived in what is now North America, having fun with dinosaurs and enjoying the scenery. Imagine Kosmoceratops attending a dinosaur masquerade —its head frills would be the coolest masks!

Did You Know?

Kosmoceratops had 15 horns on its head, more than any other dinosaur, making it one of the fanciest and most unique-looking dinos ever discovered!

Kotasaurus

Pronunciation: koh-tuh-SAWR-us

Kotasaurus was a long-necked dinosaur with a fancy name that sounded like a dinosaur explorer's map. This dinosaur was a plant-eater, using its long neck to reach high up in the trees for yummy leaves. Kotasaurus was a dinosaur gardener who helped to keep the dinosaur forest green and beautiful.

It walked on four legs with a long tail. Imagine Kotasaurus giving dino friends a tour from its high vantage point – it would be a dino travel guide!

Did You Know?
Kotasaurus is one of the earliest-known sauropods from India, giving scientists clues about the ancient forests of the region!

Khaan

Pronunciation: kahn

Khaan was a small dinosaur with a cool name that sounded like a warrior from ancient times—a real dinosaur hero! This dino was a mini T. rex with sharp teeth and a fierce attitude. Khaan was a fast runner and might have hunted for smaller animals and insects.

It was found in Mongolia, taking us on a journey to the dino world in far-off places. Imagine Khaan leading a dino parade – it would be the brave dino leader with a big roar!

Did You Know?
Khaan had a unique crest on its head, which may have been used to communicate with other dinosaurs or attract attention—like a built-in dino hat!

Lambeosaurus

Pronunciation: LAM-bee-oh-SAWR-us

Lambeosaurus had a cool-looking crest on its head as if it were wearing a dino or a fancy hat! This dino was a plant-eater, using its duck-like beak to munch on leaves and green goodies. Lambeosaurus might have used its crest to make loud honking sounds.

It lived a long time ago in what is now North America, hanging out with other dinosaurs. Imagine Lambeosaurus in a dinosaur concert—it would be the star soloist with its dinosaur trumpet!

Did You Know?
Lambeosaurus had a unique crest shaped like a hatchet or axe, and scientists think each Lambeosaurus had a slightly different crest shape, making each one unique!

Leaellynasaura

Pronunciation: lee-ELL-in-ah-SAW-rah

Leaellynasaura had a name that sounded like a dino friend's name – a buddy you'd want to have in your dino gang! This dino lived in cold places and loved to explore the freezing dino wilderness. Leaellynasaura had big eyes, which helped it see in the dark during the long polar nights.

It was a small dino and might have had fluffy feathers to stay warm – a dino with a cozy coat! Imagine Leaellynasaura leading a dino campfire storytelling session – its stories would be full of frozen dino adventures!

Did You Know?
Leaellynasaura lived in Antarctica's ancient forests, which used to be warmer than today, though still chilly enough for this dino to have some unique winter survival skills!

Leptoceratops

Pronunciation: lep-toh-SAIR-uh-tops

Leptoceratops had a fancy name that means "slim-horned face," and it looked like a dinosaur unicorn with its small horn! This dino was a plant-eater, using its sharp beak for munching on leaves and plants. Leptoceratops had a frill on its head that might have protected its neck.

It walked on four legs and had sturdy hooves. Imagine Leptoceratops in a dino talent show – it would be the charming dino magician with its dino unicorn horn!

Did You Know?
Leptoceratops had strong jaw muscles, allowing them to chew tough plants and fibrous leaves, perfect for a dino that needed lots of energy for its mountain adventures!

Liopleurodon

Pronunciation: lye-oh-PLOOR-oh-don

Liopleurodon was a super-long ocean dino with a name that sounded like a dino rockstar. This dino was a swimming superstar, with flippers that helped it glide through the dino waters. Liopleurodon had sharp teeth and could munch on large prey.

It lived long ago in the oceans, having dino adventures beneath the waves. Imagine Liopleurodon in a dino diving contest – it would be the champion with its excellent dino swimming skills!

Did You Know?
Liopleurodon could smell underwater! It had a keen sense of smell that helped it track prey, even from far away, like an underwater dino detective!

Maiasaura

Pronunciation: my-uh-SAWR-uh

Maiasaura's name means "good mother lizard"— a dinosaur supermom taking care of its dinosaur babies! This dinosaur was a plant-eater and might have lived in big groups. Maiasaura also built nests for its eggs, protecting them and keeping them warm with a duck-like beak and a nurturing spirit.

Imagine Maiasaura hosting a dino playdate – it would be the best dino babysitter for all its dino friends' babies!

Did You Know?
Maiasaura babies stayed in the nest and were cared for by their parents until they were big enough to walk, making it one of the first dinosaurs known for family care!

Megalosaurus

Pronunciation: meg-uh-lo-SAWR-us

Megalosaurus, a fearsome name that means "great lizard," was like a dinosaur king of its time! This dino was a meat-eater with sharp teeth, hunting for prey with its powerful jaws. Megalosaurus walked on two legs and had strong claws that could grip its dino meals.

It lived a long time ago in what's now Europe, ruling the dinosaur kingdom with its might. Imagine Megalosaurus in a dinosaur costume party—it would be the king of the jungle with its dinosaur crown!

Did You Know?
Megalosaurus was the first dinosaur ever named by scientists back in 1824, making it the original "discovered" dinosaur!

Microraptor

Pronunciation: MY-kroh-rap-tor

Microraptor was a small dino with a name that sounded like a dino superhero. This dino had feathers and could glide, similar to a dino flying ace. Microraptors had sharp claws and teeth, making them dino hunters in the trees.

It might have had colorful feathers that soared through the sky. Imagine Microraptor performing in a dino aerial show – it would be the star performer with its dino acrobatics!

Did You Know?

Microraptor had feathers on all four of their limbs, giving them a unique "four-winged" appearance that helped them glide from tree to tree!

Muttaburrasaurus

Pronunciation: mutt-uh-burr-uh-SAWR-us

Muttaburrasaurus's name sounded like a dance move – a dino boogie woogie in action! This dino was a plant-eater, using its strong jaw to munch on leaves and plants. Muttaburrasaurus had a bony frill on its head as its dino accessory.

It lived in what's now Australia, exploring the dinosaur world down under. Imagine Muttaburrasaurus leading a dinosaur parade—it would be the dancing dinosaur leader with its groovy moves!

Did You Know?
Muttaburrasaurus had a large, hollow nose that may have let it make loud, trumpet-like sounds, perfect for dino parades!

Nanotyrannus

Pronunciation: nan-oh-tie-RAN-us

Nanotyrannus was a mini T. rex – a small but feisty dino with a big attitude! This dino had sharp teeth for munching on meat and might have been a fast runner. Nanotyrannus was a bit like a dino ninja, sneaking up on its prey with its small size.

Scientists are still figuring out if Nanotyrannus was its dino species or just a young T. rex. Imagine Nanotyrannus in a dino race – it would be the small but speedy champion of the track!

Did You Know?
Nanotyrannus had large eye sockets, which may have given it excellent eyesight, perfect for spotting prey from a distance!

Nodosaurus

Pronunciation: NOH-doh-SAWR-us

Nodosaurus was a dino covered in armor that was always ready for a battle! This dino was a plant-eater, using its beak to munch on leaves and plants. Nodosaurus had spikes on its tail that could scare away other dinos.

It walked on four legs and was a bit like a dino tank, moving slowly and steadily. Imagine Nodosaurus in a dino parade – it would be the dino bodyguard, protecting its fellow dinos!

Did You Know?
Nodosaurus didn't have a clubbed tail like some other armored dinos but relied on its tough, bumpy armor alone to protect it from predators.

Nothronychus

Pronunciation: noh-THRON-ih-kus

Nothronychus's name sounded like a dino magician – a dino performer with a surprise up its sleeve! This dino was a plant-eater, using its beak to snip off leaves and branches. Nothronychus had long claws on its hands that it might have used for grabbing food.

It might have had colorful feathers, like a dino artist's palette of vibrant colors. Imagine Nothronychus putting on a dino magic show – it would be the dino illusionist with its enchanting tricks!

Did You Know?
Nothronychus is one of the only dinosaurs found in North America with big claws designed for reaching plants, making it a rare dino "veggie picker" in a land of fierce meat-eaters!

Nyasasaurus

Pronunciation: nye-uh-sa-SAWR-us

Nyasasaurus was one of the earliest dinos, like a dino pioneer from the ancient dino times. This small dino walked on two legs, showing the way for another bipedal dino. Nyasasaurus had sharp teeth, probably munching on insects and small critters.

It lived a long time ago in what's now Africa, leaving footprints for us to discover. Imagine Nyasasaurus as a dinosaur explorer—it would be the brave adventurer discovering the dinosaur world for the first time!

Did You Know?

Nyasasaurus might be the oldest known dinosaur or close relative, possibly dating back over 240 million years, making it one of the first to begin the incredible age of dinosaurs!

Ornithomimus

Pronunciation: or-NITH-oh-MY-mus

Ornithomimus's name sounded like a dino speedster with long legs! This dino was a fast runner. Ornithomimus had a toothless beak, probably used for grabbing dino snacks quickly.
It might have traveled in groups, running together in a dinosaur relay race.

Imagine Ornithomimus in a dinosaur race—it would be the dinosaur champion zooming past the finish line!

Did You Know?
Ornithomimus had feathers on its arms and tail, giving it a bird-like appearance and possibly keeping it warm—making it look like a running, feathered dino athlete!

Oviraptor

Pronunciation: OH-vi-rap-tor

Oviraptor had a sneaky name that meant "egg thief," but it wasn't stealing eggs – it was taking care of its own dino nest! This dino was a meat-eater, munching on small animals and insects, not eggs from other dinosaurs. Oviraptor had a beak that it might have used to crack open dino snacks.

It might have had feathers and looked like a dino-bird, not a dino thief! Imagine Oviraptor in a dino parenting class – it would be the attentive dino parent learning all the tricks of the trade!

Did You Know?
Fossils of Oviraptor have been found sitting on top of its eggs, showing it was a protective parent keeping its babies safe!

Ouranosaurus

Pronunciation: oo-RAH-noh-SAW-rus

Ouranosaurus was a dino with a built-in umbrella – it had a sail on its back to help it stay cool! This dino was a plant-eater, using its beak to munch on leaves and plants. Ouranosaurus might have used its sail to control its body temperature.

It walked on two legs or four legs, like a dino gymnast showing off different moves. Imagine Ouranosaurus at a dino pool party – its sail would be the coolest pool float!

Did You Know?
Ouranosaurus had wide, flat teeth designed for grinding tough plants, making it one of the best plant-chomping dinos around!

Ozraptor

Pronunciation: Os-rap-tor

Ozraptor's name sounds like it's from a dino adventure – a dino explorer from the land of Oz! This dino was a meat-eater, using its sharp teeth to catch prey. Ozraptor was small and walked on two legs.

It might have been a speedy runner chasing after dino snacks. Imagine Ozraptor in a dino treasure hunt – it would be the dino adventurer searching for hidden dino gems!

Did You Know?
Ozraptor fossils were found in Australia, making it one of the oldest-known dinosaurs from down under and adding to the history of ancient Australian wildlife!

Parasaurolophus

Pronunciation: par-uh-SORE-uh-lo-fus

Parasaurolophus had a cool crest on its head that looked like a dino trumpet – it could make dino music! This dino was a plant-eater, using its beak to munch on leaves and delicious greens.

Parasaurolophus walked on two legs or four legs and might have traveled in groups. It could honk or hoot through its crest, talking to other dinos in a dino phone. Imagine Parasaurolophus in a dino band – it would be the star musician with its unique dino trumpet solo!

Did You Know?

The crest of Parasaurolophus may have acted as a natural "echo chamber," amplifying its calls to be heard from miles away—perfect for dino long-distance chats!

Pachycephalosaurus

Pronunciation: pak-ee-SEF-uh-lo-sawr-us

Pachycephalosaurus had a hard head that might have used it for dino battles! This dino was a plant-eater, using its beak to nibble on leaves and plants. Pachycephalosaurus walked on two legs with a dome-shaped head.

It might have head-butted other dinos, like a dino wrestler showing off its moves. Imagine Pachycephalosaurus in a dino wrestling match – it would be the formidable dino contender with its head-thumping skills!

Did You Know?

Pachycephalosaurus had a skull up to 10 inches thick, making it one of the thickest-headed dinosaurs—like a built-in dino helmet for extra protection!

Protoceratops

Pronunciation: pro-toh-SAIR-uh-tops

Protoceratops had a name that means "first horned face" - a dino with a fashionable dino accessory! This dino was a plant-eater, using its beak to munch on leaves and tasty dino snacks.

Protoceratops had a tiny frill on its head that might have protected its neck, like a dino shield.

It might have lived in groups, hanging out with other dinosaurs and having fun. Imagine Protoceratops in a dinosaur costume parade—it would be the stylish dinosaur with its fabulous frills!

Did You Know?

Protoceratops fossils were often found with nests of eggs nearby, leading scientists to believe it was one of the earliest dinosaurs to care for its young!

Pteranodon

Pronunciation: teh-RAN-oh-don

Pteranodon was a flying dino pilot soaring through the skies with its big wings! This dino had a long beak and was like a dino fisher, catching fish from the dino waters. Pteranodon had no feathers – it had leathery wings

It lived during the time of the dinos but wasn't a true dino – it was a reptile that could fly. Imagine Pteranodon in a dino air show – it would be the dino aviator performing daring stunts in the sky!

Did You Know?
Pteranodon had a wingspan of up to 33 feet, about as wide as a small airplane, making it one of the largest flying reptiles of its time!

Qantassaurus

Pronunciation: KWAN-tuh-SAWR-us

Qantassaurus had a fun name that sounded like a dino pilot. This dino was a plant-eater, using its beak to munch on leaves and greens. Qantassaurus walked on two legs with a long tail.

It was found in what is now Australia, exploring the dinosaur world down under. Imagine Qantassaurus piloting a dinosaur plane—it would be the dinosaur captain leading the dinosaur travel adventure!

Did You Know?
Qantassaurus lived in cooler climates, and some scientists think it may have grown a thicker coat of feathers or down to stay warm during the colder seasons!

Qingxiusaurus

Pronunciation: ching-shee-uh-SAWR-us

Qingxiusaurus's name sounded like a dino artist – a dino painter creating masterpieces! This dino was a plant-eater, using its beak to snip off leaves and tasty dino treats. Qingxiusaurus had a long tail and walked on two legs.

It lived in what's now China, exploring the dinosaur world in a faraway land. Imagine Qingxiu Saurus in a dinosaur art gallery —it would be the dinosaur artist showcasing its colorful creations!

Did You Know?

Qingxiusaurus was a close relative of other long-necked sauropods and had a super long neck, perfect for reaching high leaves in the ancient forests of China!

Quaesitosaurus

Pronunciation: kway-see-toh-SAWR-us

Quaesitosaurus had a name that sounded like a dino puzzle – a dino mystery waiting to be solved! This dino was a plant-eater, using its beak to munch on leaves and delicious dino snacks.

Quaesitosaurus had a long neck and walked on four legs. It lived a long time ago in what's now North America, leaving dinosaur clues for us to discover. Imagine Quaesitosaurus as a dinosaur detective—it would be the brave investigator piecing together the dinosaur puzzle!

Did You Know?
Quaesitosaurus had a long neck but a relatively short tail for a sauropod, giving it a unique body shape compared to other long-necked dinos!

Quilmesaurus

Pronunciation: KEEL-meh-sawr-us

Quilmesaurus had a fun name that sounds like a dino dance –doing the Quilmesaurus boogie! This dino was a meat-eater, using its sharp teeth to munch on prey. Quilmesaurus walked on two legs and had long legs, acting as a dino runner racing through the dino world.

It might have been a fast dino, chasing after dino snacks with its quick steps. Imagine Quilmesaurus in a dino dance-off – it would be the dino dancer with the grooviest moves!

Did You Know?
Quilmesaurus was discovered in Argentina, and its name comes from the indigenous Quilmes people of that region, honoring their history!

Raptors

Pronunciation: RAP-tors

Raptors were like dino ninjas, sneaking up on their prey with their sharp claws and speedy moves! These dinos were clever hunters working in a dino team to catch their meals. The Raptors had big brains and were one of the smartest dinos.

They had feathers, making them look dino birds with fierce attitudes. Imagine raptors in a dino game of hide-and-seek – they would be the stealthy champions of the hunt!

Did You Know?
Raptors had a unique, curved "sickle claw" on each foot, which they could raise off the ground to keep it sharp for when they needed it—like a hidden dino weapon!

Rajasaurus

Pronunciation: RAH-jah-sawr-us

Rajasaurus had a name that means "princely lizard" – a dino with royal dino manners! This dino was a meat-eater, using its sharp teeth to catch prey. Rajasaurus had bumpy skin on its snout, making it look like a dino crown.

It lived in what is now India, ruling the dinosaur lands with its majestic presence. Imagine Rajasaurus hosting a dinosaur feast—it would be the noble host at a dinosaur banquet!

Did You Know?
Rajasaurus had a unique single horn on its forehead, adding to its "royal" look and setting it apart from other meat-eating dinos!

Rhabdodon

Pronunciation: RAB-do-don

Rhabdodon had a fancy name that sounded like a dino scientist – a dino expert in a lab coat! This dino was a plant-eater, using its beak to munch on leaves and tasty dino greens. Rhabdodon had a tall neck and a dino giraffe reaching for the yummiest dino snacks.

It walked on two legs or four legs, and it's a dino gymnast showing off its moves. Imagine Rhabdodon in a dino science fair – it would be the brainy dino showcasing its knowledge of dino plants!

Did You Know?

Rhabdodon fossils have been found on islands in what's now Europe, and it may have evolved to be smaller than mainland dinosaurs, adapting to its island life!

Rugops

Pronunciation: ROO-gops

Rugops had a name that sounded like it was from a dino rug store – like a dino with soft, cozy fur! This dino was a meat-eater, using its sharp teeth to catch prey like a dino hunter. Rugops had a bumpy head and snout, making it look a bit like a dino bulldozer.

It lived in what's now Africa, exploring the dinosaur world in far-off lands. Imagine Rugops in a dinosaur fur fashion show—it would be the stylish dinosaur with its fabulous fur coat!

Did You Know?
Rugops had a snout full of tiny holes, which may have helped it sense smells better, or even had blood vessels to help it cool down, like a natural dino air-conditioner!

Stegosaurus

Pronunciation: STEG-oh-sawr-us

Stegosaurus had spiky plates on its back like a dino superhero's armor – ready for dino action! This dino was a plant-eater, using its tiny head and beak to munch on leaves and tasty greens. Stegosaurus had a tail with big spikes that it might have used to fend off dino bullies.

It walked on four sturdy legs and had a small brain, but it was still a dino star! Imagine Stegosaurus in a dino shield competition – its spiky plates would be the dino champion's shield!

Did You Know?
Stegosaurus's back plates might have changed color to help it communicate with other dinos, making it one of the flashiest dinos around!

Spinosaurus

Pronunciation: SPY-no-sawr-us

Spinosaurus was like a dino fisher, living near water and using its long jaws to catch fish from the dino rivers. This dino was a meat-eater with sharp teeth, hunting for land and aquatic prey. Spinosaurus had a sail on its back, which might have helped it swim and regulate its temperature.

It lived during the time of the dinosaurs, ruling as one of the biggest predators—a true dinosaur king! Imagine Spinosaurus as a dinosaur lifeguard—it would be the aquatic protector of the dinosaur waters!

Did You Know?
Spinosaurus was the first-known dinosaur adapted for swimming, with webbed feet that helped it paddle through water like a prehistoric crocodile!

Stegocera

Pronunciation: steg-oh-SAIR-us

Stegoceras had a name that sounded like a dino race car – a speedy dino ready to zoom through the dino world! This dino had a bony dome on its head that might have been used for head-butting. Stegoceras had a small body and long legs, making him a dino sprinter in the Olympics.

It lived during the time of the dinos, showing off its head-butting moves to other dinos. Imagine Stegoceras in a dino head-butting contest – it would be the dino champion with its dino racing helmet!

Did You Know?
The dome on Stegoceras's head was filled with a honeycomb-like structure, making it both light and strong—perfect for head-butting matches without getting hurt!

Styracosaurus

Pronunciation: sty-RACK-oh-sawr-us

Styracosaurus had a name that sounded like a dino superhero – a dino with powerful strength! This dino had a frill on its head with big spikes that made it look fierce and cool. Styracosaurus had a beak similar to a dino parrot, using it to munch on plants and leaves.

It lived during the time of the dinosaurs and might have used its frill for defense and showing off. Imagine Styracosaurus in a dinosaur frill fashion show—it would be the dinosaur model with the most stylish spikes!

Did You Know?
Styracosaurus had six large spikes on its frill, which could grow up to two feet long—making it one of the most intimidating-looking plant-eaters around!

Triceratops

Pronunciation: tri-sera-taps

Triceratops had a cool name that sounded like a dino superhero – a dino defender of the land! This dino had three big horns on its face, like a dino with a built-in crown. Triceratops was a plant-eater, using its sharp beak to munch on leaves and dino veggies.

It had a frill on its head that might have been for show and protection. Imagine Triceratops in a dino parade – it would be the horned dino leading the way with its royal presence!

Did You Know?

Triceratops's frill was likely used for protection during battles, but it might also have been a way to attract mates—making it both a dino shield and a dino fashion statement!

Tyrannosaurus Rex

Pronunciation: tie-RAN-oh-SAWR-us REKS

Tyrannosaurus Rex was a dinosaur celebrity—a big, powerful dinosaur that everyone wanted to know about! This dino was a meat-eater with sharp teeth, hunting for prey with its strong jaws. T. rex had tiny arms but big legs, like a dinosaur with a powerful sprint.

It lived during the time of the dinos and might have been one of the fiercest predators. Imagine T. rex on a dino talk show – it would be the dino star sharing its tales of dino fame!

Did You Know?

T. rex had the strongest bite force of any land animal that ever lived, capable of crushing bones with jaws that exerted over 12,000 pounds of pressure!

Troodon

Pronunciation: TROH-oh-don

Troodon's name sounds like a dinosaur scientist— a dino with glasses and a lab coat! This dino was smart and might have been one of the cleverest dinosaurs. Troodon had a sharp claw on its toe that it might have used for hunting.

It walked on two legs and might have been a night owl – an active dino during the dark. Imagine Troodon as a dino inventor – it would be the dino creating amazing dino gadgets with its brilliant mind!

Did You Know?

Troodon had one of the largest brain-to-body ratios of any dinosaur, making it one of the smartest creatures of the Mesozoic era, like the "Einstein" of the dinosaur world!

Therizinosaurus

Pronunciation: ther-uh-ZIN-oh-sawr-us

Therizinosaurus had a long name that sounded like a mystery dino— a dino with secrets waiting to be discovered! This dino was a plant-eater, using its long claws to reach for leaves high up in the trees. It had a long neck and was a bit of a dinosaur giraffe with claws.

It might have lived in Mongolia, exploring the dino world in far-off lands. Imagine Therizinosaurus in a dino tree-climbing contest – it would be the dino winner with its amazing claw skills!

Did You Know?
Therizinosaurus had claws that could grow up to 3 feet long, making them the longest claws of any known land animal—like nature's ultimate tree climber!

Unenlagia

Pronunciation: oo-NEN-lah-gee-ah

Unenlagia's name sounded like a dino dance – a dino twirling and spinning in the dino spotlight! This dino had feathers and might have been a dino-bird with a twist. Unenlagia had sharp claws and might have used them to catch prey.

It lived in what's now South America, exploring the dinosaur world in a faraway land. Imagine Unenlagia in a dinosaur dance-off—it would be the dinosaur dancer with the most graceful moves!

Did You Know?
Unenlagia was part of a group of feathered dinosaurs that could glide or possibly even fly short distances, making it one of the earliest flying-like dinosaurs!

Utahceratops

Pronunciation: you-TAH-cer-a-tops

Utahceratops' name sounded like a dino adventure. This dino had big horns on its face and a frill that made it look tough and cool. Utahceratops was a plant-eater, using its beak to munch on leaves and tasty dino snacks.

It lived during the dinosaur era and might have hung out with other plant-eating dinosaurs. Imagine Utahceratops as a dinosaur tour guide—it would be the dinosaur expert showing us around the dinosaur terrain!

Did You Know?
Utahceratops was a close relative of the famous Triceratops and had similar features, but it lived millions of years earlier in what is now Utah!

Udanoceratops

Pronunciation: oo-DAH-noh-SAWR-us

Udanoceratops had a fancy name that sounded like a dinosaur treasure. This dino had a frill on its head with little horns that made it look unique and stylish. Udanoceratops was a plant-eater, using its beak to munch on leaves and delicious dinosaur greens.

It lived during the dinosaur era and might have been part of a dinosaur herd. Imagine Udanoceratops as a dinosaur fashion model—with its fabulous frill and horns, and it would be the trendsetter!

Did You Know?
Udanoceratops was discovered in Mongolia, and it was one of the earliest ceratopsians, a group of horned dinosaurs that later became famous for their large frills and horns!

Ugrunaaluk

Pronunciation: oo-GROO-nuh-ah-look

Ugrunaaluk has a name that sounds like a dinosaur language – a dinosaur with its own dinosaur alphabet! This dinosaur was a plant-eater, using its beak to munch on leaves and tasty dinosaur snacks. It was found in what's now Alaska, exploring the dinosaur world in the cold dinosaur lands.

Like a dino family, it might have lived in groups and laid eggs in nests. Imagine Ugrunaaluk in a dino storytelling circle – it would be the dino bard sharing epic tales of the dino past!

Did You Know?
Ugrunaaluk is one of the earliest dinosaurs known to have lived in the Arctic, showing that even in the coldest environments, dinosaurs could survive and thrive!

Valdosaurus

Pronunciation: VAL-doh-sawr-us

Valdosaurus's name sounds like a dinosaur explorer— a dinosaur discovering hidden dinosaur treasures! This dinosaur was a plant-eater, using its beak to munch on leaves and tasty dinosaur greens. It had a long neck and walked on four legs.

It lived during the time of the dinos and might have hung out with other plant-eating dinos. Imagine Valdosaurus as a dino archaeologist – it would be the dino digging up ancient dino artifacts!

Did You Know?

Valdosaurus was one of the earliest known herbivores in the sauropod group, and its fossils were found in what is now Europe, helping scientists learn more about early plant-eating dinosaurs!

Vectidraco

Pronunciation: VEK-ti-DRAY-ko

Vectidraco's name sounded like a dino superhero. This dino was a flier dino pilot gliding above the dino lands. Vectidraco had wings with a small wingspan, making it a bit similar to a dino plane.

It lived a long time ago in what's now the Isle of Wight, exploring the dinosaur world in a distant place. Imagine Vectidraco in a dinosaur air show—it would be the dinosaur aviator performing daring tricks in the dinosaur clouds!

Did You Know?
Vectidraco was one of the smallest pterosaurs, with a wingspan of just over 3 feet, making it a compact and agile flier compared to other flying reptiles of its time!

Velociraptor

Pronunciation: veh-LOH-sih-rap-tor

Velociraptor had a name that sounded like a dino racer – a speedy dino zipping through the dino world! This dino was an intelligent hunter, using its sharp claws and quick moves to catch its dino prey. Velociraptors had feathers, making them a bit of a dino-bird of prey.

It was small and might have traveled with dino wolf pack groups. Imagine Velociraptor in a dino chase – it would be the dino champion in a high-speed race!

Did You Know?
Velociraptor was likely covered in feathers, and although it's often shown as much larger in movies, it was actually about the size of a turkey, making it a surprisingly speedy and agile hunter!

Vulcanodon

Pronunciation: VUL-kan-oh-don

Vulcanodon's name sounded like a dino hero – a dino saving the day with its dino powers! This dino was a plant-eater, using its long neck to reach for leaves high up in the trees. Vulcanodon walked on four legs and might have been a bit of a dino giraffe.

It lived long ago in what's now Africa, exploring the dino world in far-off lands. Imagine Vulcanodon as a dino firefighter – the dino using its long neck to spray water on dino fires!

Did You Know?
Vulcanodon is one of the earliest known sauropods, and even though it was a plant-eater, its large size helped it defend itself from predators in its ancient world!

Wannanosaurus

Pronunciation: WAH-nah-noh-SAWR-us

Wannanosaurus's name sounded like a dino cheer – a dino with a shout of dino joy! This dino was a plant-eater, using its beak to munch on leaves and tasty dino snacks. Wannanosaurus had a short neck and walked on two legs.

It lived in what's now China, exploring the dinosaur world in a faraway land. Imagine Wannanosaurus in a dinosaur dance-off—it would be the dinosaur dancer with the most energetic moves!

Did You Know?
Wannanosaurus was one of the smallest dinosaurs of its time, measuring only about 3 feet long—perfect for quick, dino dance moves!

Walgettosuchus

Pronunciation: WALL-get-oh-SOO-kus

Walgettosuchus's name sounded like a dino adventure – an explorer from a far-off dino land! This dino was a meat-eater, using its sharp teeth to catch prey. Walgettosuchus walked on two legs and had a long tail on a daring quest.

It might have lived in what's now Australia, discovering the dinosaur world down under. Imagine Walgettosuchus as a dinosaur explorer—it would be the brave adventurer finding new dinosaur wonders!

Did You Know?
Walgettosuchus is one of the rarest known dinosaurs from Australia, offering important clues about the types of meat-eating dinosaurs that roamed the Southern Hemisphere during the Cretaceous period!

Wintonotitan

Pronunciation: WIN-toh-noh-TIE-tan

Wintonotitan had a name that sounded like a dino champion – a dino winning big in the dino games! This dino was a plant-eater, using its long neck to reach for leaves and tasty dino greens.

Wintonotitan had a small head and a big body but had a friendly smile. It lived during the dinosaur era and might have traveled in groups. Imagine Wintonotitan in a dinosaur sports event—it would be the dinosaur athlete going for the gold in dinosaur games!

Did You Know?
Wintonotitan was discovered in Australia, and it's part of the sauropod group, which means it was one of the giant long-necked dinosaurs that roamed the Earth in the late Cretaceous period!

Wuerhosaurus

Pronunciation: WUR-ho-SAW-rus

Wuerhosaurus's name sounded like a dino puzzle – a mystery waiting to be solved! This dino was a plant-eater, using its beak to munch on leaves and delicious dino snacks. Wuerhosaurus had spikes on its back and sides.

It lived during the time of the dinos and might have used its spikes for protection. Imagine Wuerhosaurus as a dino detective – it would be the dino solving dino mysteries with its spiky clues!

Did You Know?

Wuerhosaurus had large spines that protruded from its back, making it one of the few plant-eating dinosaurs that used sharp features for defense, similar to the way modern-day hedgehogs use their quills!

Xenoceratops

Pronunciation: zen-oh-SER-uh-tops

Xenoceratops' name sounded like a mystery. This dino had frills on its head with bumps that made it look unique and extraordinary. Xenoceratops was a plant-eater, using its beak to munch on leaves and dino snacks.

It lived during the dinosaur era, adding its own splash of color to the dinosaur world. Imagine Xenoceratops as a dinosaur artist—it would be the dinosaur creating fantastic dinosaur designs with its frills!

Did You Know?
Xenoceratops was one of the earliest ceratopsians to be discovered in North America, and its unique frill structure helped scientists understand how ceratopsids evolved!

Xenotarsosaurus

Pronunciation: zen-oh-TAR-soh-SAWR-us

Xenotarsosaurus's name sounded like a dino puzzle – a dino riddle to solve in the dino lands! This dino was a meat-eater, using its sharp teeth to catch prey. Xenotarsosaurus walked on two legs and might have been a speedy runner.

It lived in what's now Argentina, exploring the dinosaur world in a distant dinosaur land. Imagine Xenotarsosaurus as a dinosaur detective—it would be the dinosaur unraveling dinosaur mysteries with its sharp mind!

Did You Know?
Xenotarsosaurus had a unique foot structure with special adaptations for running, allowing it to chase down prey with incredible speed for its time!

Xiaotingia

Pronunciation: SHIAO-ting-ee-ah

Xiaotingia had a name that sounded like a dinosaur songbird with a musical touch! This dinosaur had feathers and might have been a bit of a dinosaur bird with a dinosaur twist. Xiaotingia was small and walked on two legs, a dinosaur dancer that grooved to the rhythm!

It lived during the time of the dinos and might have added color to the dino skies. Imagine Xiaotingia in a dino concert – it would be the dino musician playing sweet tunes with its feathers!

Did You Know?
Xiaotingia is considered one of the earliest known birds or bird-like dinosaurs, giving scientists important clues about the evolution of flight in dinosaurs!

Xixiasaurus

Pronunciation: shee-SHEE-ah-sawr-us

Xixiasaurus's name sounded like a dino explorer discovering treasures! This dino was a plant-eater, using its beak to munch on leaves and tasty dino greens. Xixiasaurus walked on two legs and might have been a dino gymnast.

It lived during the time of the dinos and might have traveled in herds with other dinos. Imagine Xixiasaurus as a dino tour guide – it would be the dino showing us around the dino landscapes!

Did You Know?
Xixiasaurus was discovered in China, and its fossils suggest that it was one of the earliest theropods to walk on two legs, helping scientists understand how small theropods evolved!

Yangchuanosaurus

Pronunciation: YAHNG-chwa-noh-SAWR-us

Yangchuanosaurus's name sounded like a dino king – a ruler of the dino lands! This dino was a meat eater, using its sharp teeth to catch prey. Yangchuanosaurus walked on two legs and might have been a dino wolf.

It lived during the dinosaur era and might have been one of the top predators. Imagine Yangchuanosaurus as a dinosaur king—it would be the majestic ruler of the dinosaur kingdom!

Did You Know?

Yangchuanosaurus had a long, powerful tail and sharp claws, making it a fast and efficient hunter in its environment, similar to how modern predators hunt in packs!

Yaverlandia

Pronunciation: YAH-ver-lan-dee-ah

Yaverlandia's name sounded like a dino playground. This dino was a plant-eater, using its beak to munch on leaves and delicious dino snacks. Yaverlandia walked on two legs and might have been a dino jumper.

It lived during the time of the dinos and might have explored the dino coasts. Imagine Yaverlandia at a dino beach party – it would be the fun-loving dino splashing around in the dino waves!

Did You Know?

Yaverlandia lived in what is now Europe, and it had a unique body structure, with a long tail and light build that made it a quick and agile runner!

Yandusaurus

Pronunciation: YAN-doo-sawr-us

Yandusaurus's name sounded as a dino explorer. This dino was a plant-eater, using its beak to munch on leaves and tasty dino snacks. Yandusaurus walked on two legs and might have been a bit like a dino walker.

It lived in what's now China, exploring the dinosaur world in a faraway place. Imagine Yandusaurus as a dinosaur adventurer—it would be the brave explorer finding new dinosaur wonders!

Did You Know?

Yandusaurus is one of the earliest known theropods from China, and its fossils provide important clues about the evolution of early plant-eating dinosaurs that walked on two legs!

Yutyrannus

Pronunciation: YOU-tie-RAN-us

Yutyrannus's name sounded like a dino giant. This dino was a meat-eater, using its sharp teeth to catch prey. Yutyrannus had feathers, making it a bit of a dino-bird with a fierce look.

It lived during the time of the dinos and might have roared loudly, announcing its presence. Imagine Yutyrannus as a dino announcer – it would be the dino with the loudest, fiercest roars in the dino arena!

Did You Know?
Yutyrannus is one of the largest known feathered dinosaurs, showing that even the mighty predators of the Cretaceous could have had fluffy coverings!

Zalmoxes

Pronunciation: ZAL-mock-sess

Zalmoxes's name sounds like a dino with secret dino codes to solve! This dino was a plant-eater, using its beak to munch on leaves and dino snacks. Its bony crest made it look unique and cool.

It lived during the dinosaur era, adding its own splash of color to the dinosaur world. Imagine Zalmoxes as a dinosaur artist—it would be the dinosaur creating fantastic dinosaur designs with its bony crest!

Did You Know?

Zalmoxes lived in what is now Romania, and its bony crest could have been a display feature used to attract mates or show off to other dinosaurs—kind of like a living work of art!

Zephyrosaurus

Pronunciation: ZEF-ee-roh-SAWR-us

Zephyrosaurus has a name that sounds like a dino enjoying a windy day in the dino world! This dino was a plant-eater, using its beak to munch on leaves and tasty dino snacks.

Zephyrosaurus was small and walked on two legs. It lived during the time of the dinos and might have run around the dino forests. Imagine Zephyrosaurus as a dino kite – the dino soaring in the dino skies with its leafy tail!

Did You Know?

Zephyrosaurus was one of the early ornithopods, and its light build and small size made it incredibly fast, like a mini dino sprinter zipping through the prehistoric trees!

Zuniceratops

Pronunciation: ZOO-nee-ser-a-tops

Zuniceratops had a name that sounded like a dino explorer – who loves to discover dino treasures in the deserts! This dino had horns on its face and a frill that made it look tough and cool. Zuniceratops was a plant-eater, using its beak to munch on leaves and tasty dino greens.

It lived during the dinosaur era and might have walked around the dinosaur deserts. Imagine Zuniceratops as a dinosaur tour guide—it would be the dinosaur expert showing us around the dinosaur terrain!

Did You Know?
Zuniceratops is one of the oldest known ceratopsians to have a frill and horns, providing key insights into how the more famous horned dinosaurs, like Triceratops, evolved later on!

Zuul

Pronunciation: ZOO-l

Zuul's name sounded like a dino protector of the dino lands! This dino was a plant-eater, using its beak to munch on leaves and delicious dino snacks. Zuul also had spikes on its tail that it might have used for defense.

It lived during the time of the dinos and might have roamed the dino forests. Imagine Zuul as a dino knight – it would be the armored dino guarding the dino kingdom with its spikes!

Did You Know?
Zuul is named after the "gatekeeper of the dimension" from Ghostbusters, but it was actually a real dinosaur discovered in Montana. Its tail spikes were so large that they probably acted as a defense mechanism against predators!

Explore Your Imagination!

Unlock a treasure trove of reading and coloring books!

1. Simply open your camera app on your phone.
2. Point the mobile device at the QR code
3. This will take you directly to my author central page.

or visit (http://www.letscolorimaginations.com) to delve into a vibrant world of creativity and fun.

Don't miss out—join our mailing list through the website to receive a free 22-page activity book, or download select books as MP3 files.

Happy reading and coloring !

Check out my other books at
www.letscolorimaginations.com

Magical World of Princess Coloring Book
(For Girls ages 4-8 years)

Magical World of Princess Coloring Book
(For Girls ages 8-12 years)

Princess, Unicorn, Mermaid, Fairy and Pegasus
Activity Book (For Kids ages 4-8) Volume 1, 2, 3

Princess, Unicorn, Mermaid, Fairy and Pegasus
Activity Book (For Kids ages 8-12) Volume 1, 2, 3

I Can Be Anything I Want To Be
Inspiring Careers for Girls
(Coloring Book)

Can Be Anything I Want To Be
Inspiring Careers for Boys
(Coloring Book)

Check out my other books at
www.letscolorimaginations.com

Unicorn Coloring Book (For Kids ages 4-8)

Unicorn Coloring Book (For Kids ages 8-12)

Mermaids Coloring Book (For Kids ages 4-8)

Mermaids Coloring Book (For Kids ages 8-12)

Magical World of Fairies Coloring Book
(For Kids ages 4-8)

Magical World of Fairies Coloring Book
(For Kids ages 8-12)

Enchanted Fairy World Coloring Book
(For Kids ages 8-12 & Adults)

Illustrated books

Grown up me ,will be
Inspiring careers for kids 4-8y/o

When I grow up I want to be.
Inspiring stem careers for kid 8-11y/o

Coloring books

Chibi girls coloring book

Zendoddle Designs
for kids 8-12

Careers Coloring Book

Thank you so much for purchasing my book!

Your support as a reader is invaluable, especially as a small indie publisher. Knowing that my book may be playing a role in your child's education is incredibly fulfilling.

If you could spare a minute to leave your honest feedback on Amazon, it would be immensely helpful. Your review not only aids in improving my future work but also guides other parents in their choice of educational resources.

Once again, thank you for being a part of this journey. Your thoughts and experiences with the book are truly appreciated!

To leave feedback

1. Open up your camera app.
2. Point your mobile device at the QR code below.
3. The review page will appear in your web browser.

or you can search on Amazon for the name of the book and Elizabeth Shelby